MW00509365

The Chemistry Elemental Book of 118 Poems

The Chemistry Elemental Book of 118 Poems

Walter the Educator

Silent King Books a WhichHead Imprint

First Printing, 2023

dedicated to all the chemistry lovers

Why I created this book?

Creating a poetry book about the periodic table of elements can be a unique way of exploring scientific concepts that may seem dry and technical to some. Poetry can give a fresh perspective and bring an imaginative approach to learning about the elements. It can also help students remember chemical elements and their properties better. Moreover, it can bridge the gap between science and art and appeal to a larger audience. Overall, a poetry book about the elements within the periodic table can be an appealing and creative way to engage with science.

Walter the Educator

Hydrogen 1H

Invisible yet everywhere,
Hydrogen bonds beyond compare.
The simplest element in the book,
But its power, don't overlook.
It fuels the stars up in the sky,
And makes up water, so pure and spry.
Explosive when it meets a flame,
Its versatility, hard to tame.
From fuel cells to rocket fuel,
Hydrogen's uses, always cool.
Light as air and oh so clean,
The energy of the future, it might seem.
So let's celebrate this gas so light,
The element with atomic might.
Hydrogen, our friend in space,
We'll never forget your special place.

Helium 2He

Inert and noble, you float with ease
A gas so light, you rise with breeze
Your atomic number two, your weight is small
Your uses diverse, you do enthrall
From balloons to blimps, you lift them high
Your voice goes squeaky, when you give a try
In welding and cooling, you play a part
In MRI machines, you take the lead with heart
The sun's nuclear fusion, you play a role
In the universe, you're not so small
From stars to galaxies, you're found in space
In the periodic table, you have a place
Oh helium, you're truly unique
Your properties, a physicist's critique
A gas so scarce, yet so important
In our world, you're truly magnificent

Lithium 3Li

Lithium, Lithium, oh element of three
One of the lightest, so simple and free
In the periodic table, it stands with glee
A metal that's rare, but oh so key
A symbol of Li, it's atomic number three
In stars it was born, from fusion of three
A metal that's soft, it's silvery-white
It's reactive, it's volatile, but oh so bright
From batteries to medication, it has a role to play
Depression, bipolar, it helps in every way
It's used in nuclear reactors, and fusion too
It's an element so versatile, pure and true
Lithium, Lithium, oh element of bliss
Your atomic weight so low, your power so immense
We thank you for your service, for all that you do
A precious element, forever in our view

Beryllium 4Be

Beryllium, oh Beryllium!
A shining element, oh so premium!
With four electrons, a nucleus so small
It's properties, unique and exceptional.
It's light and strong, a metal so rare,
Found in minerals, it's beauty so fair.
It's used in aerospace and nuclear reactors,
And in X-ray windows, it's a true factor.
It's toxicity, a danger so great,
But still, it's properties we can't debate.
It's a wonder of chemistry, a sight to see,
Oh Beryllium, you'll always be a mystery.

Boron 5B

Boron, oh Boron, with five protons in tow,
You sit in Group 13, with a valence of three, you know.
Your atomic number is five, and you're a metalloid,
With a unique structure that's not easily destroyed.
In compounds, you may take on different forms,
From borates to boranes, and even boric acid storms.
You make strong bonds with carbon, and your uses are vast,
From rocket fuel to ceramics, and even nuclear blast.
But beware of your dangers, oh Boron, so keen,
For your compounds can harm, and your powder's quite mean.
Your toxicity is high, and your dust can ignite,
So handle with care, and keep Boron out of sight.

Carbon 6C

Carbon, oh carbon, so versatile and pure
The foundation of life, essential for sure
In diamonds, you shine with a brilliant light
In graphite, you glide with ease and might
Your chains form the backbone of every living cell
Your bonds create the fuels that we burn so well
From the depths of the Earth, we mine your black gold
In the sky, you absorb the heat that we hold
Your isotopes reveal our planet's age
Your dating methods unlock history's page
In nanotubes, you hold the key to change
In semiconductors, you power the range
Carbon, oh carbon, so simple and grand
Your power and potential, we cannot understand
From the tiniest atom to the largest star
Your presence is felt, near and far.

Nitrogen 7N

Nitrogen, oh nitrogen, a gas so inert,
In our atmosphere, it's the most prevalent expert.
It's the building block of life, in proteins it's found,
In DNA and RNA, it's always around.
Its triple bond is strong, and hard to break,
As a fertilizer, it's used to enhance the soil's stake.
It's a cryogenic fluid, used to freeze and store,
In the food industry, it's used more and more.
It makes up 78% of our air we breathe,
Without it, life on earth would never achieve.
From the stars it was born, in supernovae's light,
Nitrogen, oh nitrogen, such a wondrous sight.

Oxygen 8O

Oxygen, the breath of life
Without you, we'd be in strife
A gas that we can't see or taste
But vital for all life to be placed
You make up a fifth of the air we breathe
And in water, you're what we need
You're part of every molecule
That makes up life, it's no fool
You help fuel the fire of life
And in our bodies, you're rife
You give us the energy we need
To live, to dream, to plant the seed
Oxygen, you're a mystery
But we know your chemistry
You're essential for us to thrive
And without you, we couldn't survive.

Fluorine 9F

Fluorine, Fluorine, the element of cheer,
A reactive gas, abundantly clear.
Its atomic number is nine, so fine,
It's easy to find in the periodic line.
In nature, it's found in minerals so bright,
Like fluorite and topaz, a lustrous sight.
It's used in toothpaste, Teflon, and more,
A versatile element, we can't ignore.
Its electronegativity is high,
It attracts electrons like a butterfly.
A halogen, among the noble gases,
Its reactivity is what sets it apart from the masses.
Fluorine, Fluorine, so vital to life,
In our bones, it helps to prevent strife.
But if we're not careful, it can be a danger,
A toxic gas, we must not anger.
So let us appreciate Fluorine, our friend,
A vital element, on which we depend.
With its power and versatility,
Fluorine, we salute thee!

Neon 10Ne

Neon, oh Neon, you shine so bright,
A gas so noble, a source of light,
Your electrons, tightly held,
A glowing beauty, you do weld.
In tubes and signs, you make us see,
A spectrum of colors, so vividly,
From red to blue, and all between,
A rainbow of light, like we've never seen.
A rare gas, you're hard to find,
But in the air, you're not confined,
A small fraction, you do make,
A shining star, for us to take.
Oh Neon, you're a wonder to behold,
A unique element, so precious and bold,
In the darkness, you come alive,
A shining light, that will survive.

Sodium 11Na

Sodium, oh sodium, 11th in line,
A metal so reactive, it's hard to confine.
Symbol Na, from the Latin word Natrium,
In compounds and salts, it's found in a stadium.

It's soft and silvery, like a shining star,
But keep it away from water, or it'll go far.
It'll fizz and pop, and even catch fire,
A sight so amazing, you'll never tire.

In our bodies, sodium plays a vital role,
Helping with nerve impulses, it's part of the whole.
But too much of it, can lead to some trouble,
Like high blood pressure, it's best to avoid the double.

Sodium, oh sodium, you're part of our life,
In food and drinks, you help us survive.
But remember to use you, in moderation and care,
For a healthy and happy life, that's only fair.

Magnesium 12Mg

Magnesium, oh Magnesium,
A metallic element with a spark,
Its brilliance shines in daylight,
But in the dark, it truly embarks.
In flames, it burns with a passion,
A blaze of white hot light,
Its energy is truly captivating,
A magnificent, awe-inspiring sight.
In the body, it plays a role,
A vital element it is,
For muscles and bones, it's important,
A true mineral, that's the gist.
Magnesium, oh Magnesium,
A chemical element so fine,
Its properties are truly unique,
A treasure of the periodic line.

Aluminum 13Al

It's properties, truly a wondrous sight
From soda cans to airplanes in flight
Its versatility, always shines bright

Discovered in the 19th century, so new
A precious metal, so rare and few
But bauxite changed the game, who knew?
Now abundant, it's uses grew and grew

It conducts electricity with ease
And resists corrosion, if you please
Reflective, it shines like a tease
And with strength, it can surely appease

Aluminum, oh how you impress
A metal that we surely can't repress
From cans to cars, you truly bless
Our world, with your adaptiveness

Silicon 14Si

Silicon, oh Silicon, so abundant and pure
Your shining crystal structure, so perfect and secure
You're found in every corner, a vital part of our world
In electronics and solar cells, your marvels unfurled
Your atomic number is fourteen, your mass is twenty-eight
Your valence electrons make you, a semiconductor great
You bond with other elements, to make materials strong
From computer chips to glass, your usefulness prolong
Silicon, oh Silicon, you're a fascinating element
Your properties and applications, truly heaven-sent
A symbol of innovation, a cornerstone of technology
Silicon, oh Silicon, you're an integral part of our ecology.

Phosphorus 15P

Phosphorus, oh Phosphorus,
Number fifteen on the periodic chart,
A non-metal element, so luminous,
A vital player in life's grand start.
Found in DNA, in every cell,
Makes up the backbone of our genes,
In ATP, it helps us excel,
Gives us energy, like a well-oiled machine.
White, red, or black, it takes many forms,
A matchstick's tip, it ignites with ease,
A dangerous weapon, in war it performs,
A deadly poison, handle with care, please.
Phosphorus, oh Phosphorus,
A versatile element, we can't do without,
From the stars above, to the earth below us,
Its impact on life, we can't live without.

Sulfur 16S

Sulfur, the element of fire and brimstone,
A yellow hue that burns like a hot stone,
It's found in nature, in rocks and in soil,
And in the air, it can cause quite a turmoil.
In chemistry, it's used in many ways,
From making gunpowder to bright dyes that amaze,
It's in fertilizers and pharmaceuticals too,
Sulfuric acid, a powerful brew.
But sulfur has a darker side as well,
When burned, it releases a noxious smell,
And in the atmosphere, it forms acid rain,
A problem we must work to sustain.
So while sulfur may have its ups and downs,
Its importance to our world is renowned,
We must use it wisely, with care and with thought,
And keep our environment, in balance, well-fought.

Chlorine 17Cl

Chlorine, oh Chlorine, the element so green,
A powerful gas, but not to be seen,
With a pungent smell and a toxic flair,
Handle with care, or beware, beware!
Used to disinfect, to purify, to clean,
In pools and water, it keeps us serene,
In bleach and chemicals, it's always there,
A versatile element, beyond compare.
But with great power, comes great responsibility,
In wrong hands, it can cause much hostility,
Used as a weapon, it's a deadly affair,
So let's use it with caution, with utmost care.
Chlorine, oh Chlorine, an element so grand,
With its many uses, it's in high demand,
But let's not forget, its potential harm,
Let's use it wisely, with an outstretched arm.

Argon 18Ar

A noble gas, so inert and still
Argon is its name, with no desire to thrill
No color, no smell, no taste at all
In the air, it's found, but in small
Used in light bulbs, welding, and lasers
Its properties, so unique, are true amazers
In welding, it shields, to prevent oxidization
In bulbs, it glows, with no deviation
Though it's not toxic, it can cause harm
In confined spaces, it can cause alarm
Displacing oxygen, it can lead to suffocation
So use it with care, to avoid a dire situation
Argon, a gas, with many uses so bright
A cautionary tale, to use it just right.

Potassium 19K

Potassium, oh Potassium,
A metal soft as butter,
Its reactivity is fun,
But in water, beware the splutter.
From bananas to our nerves,
Potassium is vital indeed,
Without it, our muscles won't serve,
A deficiency, our bodies can't concede.
In compounds, it comes in handy,
Potassium carbonate, chloride, and more,
Its uses are plenty,
In agriculture, medicine, and industry galore.
But use it with care,
For its reaction with air can be explosive,
Treat it with respect, be aware,
Potassium, an element impressive.

Calcium 20Ca

Calcium, oh calcium, so essential you are
In bones and in teeth, you make them both strong
Your role in the body, we can't ignore
From muscle function to nerve transmission, you belong
You're found in milk, cheese, and yogurt too
Leafy greens and nuts, you're there as well
Your deficiency can cause a lot of issue
Like brittle bones and teeth, so we must dwell
On the importance of calcium, in our everyday diet
To keep our body healthy, and functioning right
Calcium, oh calcium, we need you so
To keep us strong and healthy, from head to toe

Scandium 21Sc

Scandium, oh scandium,
A rare earth metal with a glow,
Found in minerals and ores,
In the earth's crust deep below.
A silvery-white element,
With properties unique,
Strong, lightweight, and versatile,
In alloys it can speak.
Used in aerospace and racing,
For its strength and heat resistance,
Also in sports equipment,
For its lightness and persistence.
In medicine, it's a tracer,
For imaging bones and tissues,
And in lamps, it shines so bright,
For its spectral color issues.
Scandium, oh scandium,
A treasure of the periodic table,
With applications many,
For our modern world, so stable.

Titanium 22Ti

Titanium, oh titanium,
A metal strong and light,
Used in planes and spacecraft,
To soar through starry night.

Its strength is like no other,
Resistant to corrosion,
In medical implants it's used,
For its biocompatibility, a crucial decision.

From golf clubs to eyeglasses,
Its versatility is grand,
In paint, sunscreen, and food,
It's found across the land.

Oh titanium, oh titanium,
A metal of the future,
As technology advances,
Your importance will only nurture.

Vanadium 23V

Vanadium, oh Vanadium,
A metal with a name so grandium.
In nature, you're rare,
But in industry, you're everywhere.
Your strength and durability,
Are sought after with great fervor,
In steel, you play a vital role,
Making it tough and corrosion whole.
But you're not just a one-trick pony,
For in batteries, you're a key crony,
As Vanadium Redox Flow,
You store energy for a steady flow.
And in medicine, you're showing promise,
Fighting cancer with metal chelation,
Vanadium complexes are the premise,
For a new form of treatment sensation.
Oh Vanadium, you're truly unique,
In your properties and applications so chic,
May your future be bright and full of might,
As industries continue to take flight.

Chromium 24Cr

Chromium, oh Chromium,
A metal with a shine.
Used in stainless steel,
And many a design.
Its hardness and luster,
Make it perfect for plating.
And when it's in pigments,
The colors are captivating.
From car parts to kitchenware,
Chromium is everywhere.
In the tanning of leather,
And even in our air.
But beware of its compounds,
They can be toxic and harmful.
So handle with care,
And keep your safety armful.
Chromium, oh Chromium,
A versatile element indeed.
May we use it with caution,
And for good, we shall succeed.

Manganese 25Mn

In the Earth's crust, you can be found
A metal that's strong, Manganese is renowned
Used in steel to make it tough and strong
A vital element, it doesn't belong
In batteries, you also play a part
A key component, you help them start
And in medicine, you're used to treat
Diseases like anemia, a vital feat
Manganese, oh Manganese
So versatile, you lend a hand
In industry, health, and technology
Your uses go beyond, oh so many
A vital metal, we can't ignore
In so many things, you help us soar

Iron 26Fe

Iron, oh iron, how strong you are
A metal so useful, you shine like a star
In buildings and bridges, you hold them up high
And in cars and machines, you make them fly
Your rust may be feared, but your strength is admired
From swords to plows, you've never tired
Your magnetic force, like a compass, guides us true
And in our blood, you help carry oxygen through
Oh iron, you're in the earth, the stars, and beyond
A sturdy element, forever to be fond
So let us honor you, with every beat of our heart
For without you, this world would fall apart.

Cobalt 27Co

Cobalt, oh Cobalt, a metal blue,
A color so striking and bright,
You're found in Earth, so rare and true,
A lustrous metal, shining in light.
You're used in batteries, strong and true,
From phones to cars, you power them all,
You're in jet engines, turbines too,
Cobalt, you're essential, standing tall.
In medicine, you play a part,
A catalyst for healing and health,
From treating anemia in the heart,
To imaging the brain with stealth.
Cobalt, oh Cobalt, you're so versatile,
A metal of importance, we can't deny,
From industry to health, you make us smile,
Cobalt, oh Cobalt, you'll never die.

Nickel 28Ni

Nickel, oh nickel, a metal so bright
With its silvery shine, it's quite a sight
It's tough and strong, a metal so true
And its uses are many, it's not just a few

From coins to batteries, it's everywhere you see
In stainless steel, it's a crucial key
It's a catalyst in chemical reactions
And it's used in electroplating with great satisfaction

In alloys, it's a common addition
And its magnetic properties are a great attraction
It's a conductor of heat and electricity too
And in superalloys, it's a crucial ingredient through and through

Oh nickel, oh nickel, you're so versatile
And your strength and beauty are truly worthwhile
In so many industries, you play a vital role
And your importance to mankind, we cannot control.

Copper 29Cu

Copper, oh copper, a metal so red,
A conductor of heat, electricity it spreads.
A staple in coins and electrical wires,
It's antibacterial properties, a boon for medical desires.
From Bronze Age tools to modern-day art,
Copper's versatility sets it apart.
Its malleability and ductility,
Make it perfect for crafting with agility.
In architecture and roofing, it's a popular choice,
And in brewing, it gives beer its unique voice.
In the body, it aids in the formation of blood,
Copper, oh copper, its importance understood.

Zinc 30Zn

Zinc, a metal with a bluish-white hue,
A versatile element with much to do,
Used to galvanize steel and prevent corrosion,
And also in batteries, a source of power in motion.
A key component in enzymes and proteins,
Zinc is essential for the body's well-being,
From wound healing to DNA synthesis,
It plays a role in various biological processes.
Zinc oxide, a common ingredient in sunscreen,
Protects the skin from harmful ultraviolet beams,
And in cosmetics, it gives a glow and shine,
A mineral so precious, oh so divine.
Zinc, a metal so often overlooked,
But its importance cannot be overbooked,
From industry to health, it plays a critical role,
Zinc, a metal with a heart and soul.

Gallium 31Ga

Gallium, oh gallium, so rare and so bright
A metal that's soft, yet its melting point's high
With a silvery shine, at first it delights
But add heat, it melts, a unique sight
Used in semiconductors, LEDs, and more
It's a key player in the tech we adore
Gallium arsenide, a compound so grand
Makes fast transistors, with precision at hand
In medicine too, gallium finds its place
As a tracer, it helps to find the right space
To detect tumors, infections, and such
Aiding doctors, who care for us so much
Oh gallium, how versatile you are
A metal with uses, both near and far
May we continue to learn, and to explore
The wonders of gallium, forevermore.

Germanium 32Ge

Germanium, oh Germanium,
A metalloid with a story to tell,
Used in electronics and optics,
Infrared detectors and solar cells.
Its unique properties make it special,
A semiconductor with a diamond-like structure,
It's used in transistors and diodes,
A material of the future.
Its name derived from the Latin word for Germany,
But it's found in many places today,
From fiber optics to chemotherapy,
Germanium is here to stay.
Oh Germanium, you may be underrated,
But your importance cannot be denied,
A versatile element in many ways,
In science and technology, you'll always abide.

Arsenic 33As

Arsenic, oh arsenic, a metalloid of old,
Its properties are lethal, or so we are told.
A poison to humans, yet a remedy too,
In small doses, it's a cure, for what ails you.
Used in agriculture, for pest control,
In semiconductors, it plays a vital role.
In glassmaking, it gives a tint of green,
And in fireworks, it's a sight to be seen.
But beware of its dangers, for it can kill,
A toxic substance, that requires skill.
Handle it with care, and use it with caution,
For arsenic, oh arsenic, demands respect and precaution.

Selenium 34Se

Selenium, oh Selenium,
A wonder element, a true gem,
Named after the moon goddess, Selene,
A unique element, like nothing we've seen.
With its atomic number 34,
It's a metalloid, it's so much more,
In its grey form, it's a brittle solid,
But when heated, it becomes quite pliant.

Used in glassmaking and pigments,
Selenium is truly magnificent,
It's important in the body's metabolism,
And helps prevent cellular oxidation.
Selenium, oh Selenium,
A key element in the world of medicine,
From cancer prevention to thyroid health,
Selenium's benefits are vast and felt.

A vital element, a true essential,
Selenium is truly exceptional,
But like all things, it must be handled with care,
For too much of it can be quite rare.

So let us appreciate this element so,
And use it wisely, with respect and know,
That Selenium, oh Selenium,
Is truly a wonder, a true gem.

Bromine 35Br

Bromine, oh Bromine,
A halogen you are,
A red-brown liquid,
That can take you far.

You're used in dyes,
And photographic film,
Your flame-retardant properties,
Are quite the win.

But beware, oh beware,
Of your toxic gas,
Handle with care,
Or it could be your last.

From the sea you came,
In the form of salts,
Your name comes from the Greek,
For a stench that assaults.

So here's to you, Bromine,
May we use you with care,
For in the world of chemistry,
You have a place quite rare.

Krypton 36Kr

In noble gases, Krypton reigns supreme,
Its beauty lies in its rarity, like a dream,
A colorless and odorless gas, it's true,
But its properties are what make it unique and new.
Used in lighting, Krypton shines so bright,
It's stable and non-reactive, a true delight,
In lasers, it's a crucial component too,
Its uses are many, and its benefits, too.
With its full outer shell, it stands alone,
Inert and unyielding, like a throne,
But don't underestimate its power,
For in the right conditions, it can flower.
Krypton, noble and proud,
A gas that stands out from the crowd,
Its properties may not be apparent,
But in its element, it's truly inherent.

Rubidium 37Rb

Rubidium, oh Rubidium, so reactive and rare,
Your properties are unique, that's why we must beware.
You ignite in air, and water makes you hiss,
But in the right conditions, your uses are hard to miss.
In atomic clocks, your accuracy is key,
And in photocells, you help us to see.
Your salts give fireworks their red hue,
And in research, you help us through and through.
But handling you is tricky, we must take great care,
For contact with skin can cause quite a scare.
So let's appreciate your properties, but use you with caution,
For Rubidium, oh Rubidium, demands our utmost attention.

Strontium 38Sr

Strontium, a metal so bright,
In fireworks it's quite a sight,
Its red flame lights up the night,
And makes our celebrations ignite.
But this element is more than that,
It's used in medicine, imagine that,
To ease bone pain and combat,
Osteoporosis, a disease so fat.
Its properties, so unique,
Make Strontium quite a treat,
It's reactive, yet stable and meek,
And in nature, it's rare and discreet.
So let us cherish Strontium,
This element so bright and fun,
From fireworks to medicine,
Its uses are never done.

Yttrium 39Y

Yttrium, a rare earth metal with a silver shine,
Discovered in rocks, with a story quite divine.
Named after a village, its uses are quite vast,
In lasers and LEDs, it shines bright and fast.
A catalyst for chemical reactions in labs,
Yttrium's properties, scientists like to grab.
In superconductors and alloys, it plays a key role,
And in radiation therapy, it helps to make us whole.
Yttrium, oh Yttrium, we cherish you so,
For all the ways you help in science to grow.
From MRI machines to airplane parts,
You're a vital element, dear to our hearts.

Zirconium 40Zr

Zirconium, oh Zirconium,
A metal strong and true,
Used in nuclear reactors,
And spacecraft, too.
Its atomic number is forty,
And its symbol is Zr,
It's corrosion-resistant,
And shines like a star.
Zirconium dioxide,
Is a ceramic so fine,
Used in dental implants,
To make teeth align.
It's also used in alloys,
For strength and resistance,
And in the production of steel,
It adds a bit of persistence.
Oh Zirconium, oh Zirconium,
A metal with such grace,
Used in so many ways,
In this vast and varied space.

Niobium 41Nb

Niobium, oh Niobium,
Element forty-one.
A lustrous, soft, and grayish metal,
Understood by only some.
In the earth, you're not abundant,
But you're vital to our lives.
You strengthen steel and alloys,
And make them more refined.
Superconducting magnets,
MRI machines too,
All rely on Niobium,
For their work to ensue.
Niobium, oh Niobium,
So important, yet so rare.
We thank you for your service,
And for all the ways you care.

Molybdenum 42Mo

Molybdenum, oh Molybdenum
A metal, so strong and bold
In steel and alloys, you strengthen them
And make them unbreakable and whole
Your high melting point and density
Make you a valuable find
In superalloys and jet engines
You help us leave the ground behind
In enzymes, you play a role so grand
Assisting in nitrogen fixation
Without you, life could not sustain
And would suffer a great affliction
We thank you, dear Molybdenum
For your service and your care
You make our world a better place
And we are grateful that you're there.

Technetium 43Tc

Technetium, element forty-three,
A metal with a story quite unique.
Discovered in nineteen thirty-seven,
By Perrier and Segrè, heaven sent.
It's a man-made element, you see,
But still holds importance chemically.
Used in medical imaging, it's true,
Technetium-99m, a lifesaving view.
Its short half-life makes it ideal,
To diagnose and treat, to heal and feel.
From cancer to heart disease,
Technetium helps with ease.
Though scarce in Earth's crust,
In nuclear reactors it's a must.
A fission product, it's created,
But with caution, it must be treated.
Technetium, a metal with a unique tale,
From medical imaging to nuclear avail.

Ruthenium 44Ru

Ruthenium, oh Ruthenium,
A metal that's not so common,
But in the lab, it shines so bright,
And its uses are quite a sight.
In alloys, it makes them strong,
And in electronics, it belongs,
A catalyst for many reactions,
It speeds up chemical transactions.
In jewelry, it's a rare delight,
A metal that sparkles with such might,
And in space, it's used to coat,
To protect from the sun's hot rote.
Ruthenium, oh Ruthenium,
A metal with such versatility,
From solar cells to cancer drugs,
Its impact is seen in many hugs.

Rhodium 45Rh

Rhodium, rare and precious,
A metal with a shining face,
Found in ores of platinum,
A treasure in a hidden place.
Its beauty is not just skin deep,
Rhodium's properties are vast,
A catalyst for chemical reactions,
A metal that will forever last.
From catalytic converters in cars,
To treating cancer with its drugs,
Rhodium plays a crucial role,
In medicine and technology, it's a plug.
Rhodium's uses are abundant,
Its value is beyond measure,
A metal that shines like a star,
Rhodium, a true chemical treasure.

Palladium 46Pd

Palladium, oh noble metal,
Your value is so high and yet so subtle.
Used in catalytic converters to clean the air,
You're a precious element, beyond compare.
Your luster shines in jewelry and watches,
As a symbol of luxury and status.
But your true worth lies in your ability,
To help fight cancer with your versatility.
In fuel cells and electronics, you play a crucial role,
Your conductivity and durability, a wonder to behold.
From dentistry to space exploration,
Your applications are diverse without hesitation.
Palladium, you may be rare,
But you're a precious element beyond compare.
Your contributions to science and industry,
Are truly invaluable for all humanity.

Silver 47Ag

Silver, oh silver, a precious metal indeed
With a lustrous shine, it's beauty we need
From jewelry to coins, it's been widely used
A symbol of wealth, it's value never refused
In chemistry labs, it's a versatile tool
Reacting with ease, it plays by the rules
From photography to medicine, it has a role
With antimicrobial properties, it's a treasure to behold
As ions in solution, it conducts electricity
A catalyst in reactions, it's full of activity
In mirrors and glass, it's a reflective delight
Silver, oh silver, a chemistry wonder, shining bright.

Cadmium 48Cd

Cadmium, oh cadmium,
A silvery-white metal so rare.
In batteries, pigments, and coatings,
Your versatile nature is beyond compare.
But caution must be taken,
As toxicity is your dark side.
Pollution from industry and mining,
Has led to environmental divide.
But in the right hands,
You can shine like a star.
Used in nuclear reactors,
Your neutrons can take us far.
Cadmium, oh cadmium,
You may have a dark past.
But as we learn to use you wisely,
The future with you will surely last.

Indium 49In

Indium, oh Indium, with a lustrous glow
A metal that shines, like freshly fallen snow
A silvery-white hue, that's soft to the touch
Indium, oh Indium, it's simply too much
Used in touchscreens, and LED lights too
Indium tin oxide, it conducts like a pro
With its low melting point, it's easy to mold
Indium, oh Indium, it never gets old
In nuclear reactors, it can absorb neutrons
Indium, oh Indium, its uses are proven
A rare metal indeed, but not to be missed
Indium, oh Indium, it's hard to resist

Tin 50Sn

Tin, oh tin, a humble metal so pure,
In nature, it's found, without any allure,
Its beauty lies in its versatility,
From bronze to pewter, it's a metal of ability.
In cans and containers, it holds our food,
In roofs and gutters, it keeps us dry and good,
In alloys with copper, it creates bronze statues,
In mirrors and lenses, it reflects our views.
But tin has a dark side, that we must not forget,
In organotin compounds, it can be a threat,
To marine life and environment, it can cause harm,
So let's use tin wisely, and keep it away from alarm.
Tin, oh tin, a metal so dear,
May we use it with care, and keep it near,
For in its versatility and potential,
Lies the beauty of tin, so essential.

Antimony 51Sb

Antimony, oh Antimony,
A metalloid with a unique melody,
Used in flame-retardant materials,
And for medicinal purposes quite essential.
In ancient times, it was a cosmetic,
With a history that's almost mystic,
Now found in batteries and semiconductors,
As well as in pewter, an alloy that's proper.
It's used in flame-proofing textiles,
And in the production of glass quite versatile,
It's also found in bullets and batteries,
A metalloid with many abilities.
But beware, for its compounds can be toxic,
Poisonous to living beings aquatic,
So let's use it with care and caution,
Antimony, a metalloid of great proportion.

Tellurium 52Te

Tellurium, oh tellurium,
A rare and precious gift,
Your properties so unique,
Many secrets you do lift.
A metalloid with a silvery shine,
You're used in many ways,
In alloys, solar cells, and glass,
You even brighten up displays.
You're a semiconductor,
With conductivity so high,
And in thermo-electric devices,
You help convert heat to energy nigh.
But beware, dear tellurium,
In certain compounds, you may harm,
Organotelluriums can be toxic,
And cause damage to many a farm.
So let us use you wisely,
And appreciate your worth,
Tellurium, oh tellurium,
A treasure of the earth.

Iodine 53I

Iodine, oh Iodine,
Element number 53,
A nonmetal with a hue so fine,
In its pure form, a violet sheen.
Tincture of Iodine, a common sight,
Used to disinfect wounds so tight,
A drop of Lugol's solution,
Reveals the presence of starch with resolution.
Thyroid hormones, Iodine's fame,
Regulating metabolism, a crucial game,
Deficiency can lead to an enlarged gland,
But too much can cause thyrotoxicosis, oh so grand.
Iodine, oh Iodine,
A necessary element, so divine,
But use with care, my dear,
For its toxicity can bring fear.

Xenon 54Xe

Xenon, noble gas of group 18,
A colorless element, rare and serene.
Unreactive and stable, it stands alone,
Inert and aloof, a true noble stone.
Used in lamps for its bright blue light,
And in anesthesia to put us out of sight.
Xenon's unique properties make it rare,
A noble gas with characteristics beyond compare.
In space, it's found in the atmosphere,
A trace element with a presence so clear.
Xenon, inert and rare,
A noble gas with a beauty so fair.

Caesium 55Cs

Caesium, so rare and bright,
A metal that's pure and light.
Soft to touch, it melts like wax,
And reacts with water in a flash.
Its electrons move with ease,
In a state of purest peace.
And though it's hard to find,
Its uses are not confined.
In atomic clocks it's found,
Counting seconds, never bound.
And in space it helps us see,
Using gamma rays to map debris.
Caesium, a wonder element,
So versatile and elegant.
In our world it has a place,
Shining with its unique grace.

Barium 56Ba

Barium, oh barium, so shiny and bright,
A metallic element, with a lustrous light.
Soft and silvery, it's quite reactive,
In air it quickly forms an oxide protective.
Used in fireworks, to give a green hue,
Barium nitrate is what they use to do.
In medicine, it's swallowed to check the guts,
A barium meal, to locate the what's and whatnots.
It's toxic, so careful, you must be,
But in labs, it's used to test for sulphate, don't you see?
Barium, oh barium, so useful and bright,
We'll keep on using you, day and night.

Lanthanum 57La

Lanthanum, a rare earth metal,
In the periodic table, its place is settled.
Soft and malleable, it can be shaped with ease,
Its properties are unique, it's hard to tease.
Used in camera lenses, it enhances the view,
Its magnetic properties are strong, it's true.
In batteries and electric cars, it plays a part,
Reducing emissions, it's a modern-day art.
Its compounds are used in oil refining,
And in hydrogen storage, it's defining.
Lanthanum is versatile, its uses are broad,
A valuable element, it's loved and adored.

Cerium 58Ce

Cerium, oh cerium,
Rare earth and so supreme,
With an atomic number 58,
A versatile element that's truly great.
In its pure form, it's a silvery-white,
But it can oxidize to yellow, brown, or even bright,
It's used in alloys, as a catalyst too,
And in lighter flints, it helps to make the spark that's true.
In glass manufacturing, it's a vital part,
Removing impurities and lending strength to the heart,
It's also found in self-cleaning ovens and fuel cells,
Making our lives easier, it truly excels.
Cerium, oh cerium,
An element we can't ignore,
For all the ways it helps us,
We must give it applause and more.

Praseodymium 59Pr

Praseodymium, a rare earth gem,
A metal that's not quite like them,
With a greenish-yellow hue,
Its beauty shines through.
It's used in alloys and magnets,
And in glass for camera lenses,
Fuel cells and lasers, too,
Praseodymium can do.
Its atomic number is 59,
And its properties are just fine,
A soft metal that's easy to work,
Praseodymium doesn't shirk.
So let us celebrate this element,
And its contribution so excellent,
For without Praseodymium's aid,
Many industries would fade.

Neodymium 60Nd

Neodymium, oh Neodymium,
Rare earth metal, shining gem.
Magnetic power, strong and true,
In motors, headphones, and speakers too.
With its strength, it holds the key,
To technology that sets us free.
From wind turbines to electric cars,
Neodymium takes us far.
In lasers, it shines so bright,
A spectrum of colors, a dazzling sight.
From medicine to space, it's used with grace,
Neodymium, a versatile ace.
So let us honor this element,
For all the ways it has been sent,
To change our world in every way,
Neodymium, we'll always say.

Promethium 61Pm

Promethium, oh Promethium,
Rare and radioactive, you are.
First discovered in '45,
In a nuclear reactor, not afar.
Your isotopes are numerous,
And your uses, quite diverse.
From atomic batteries in space,
To measuring thickness, you disperse.
You glow a pale blue hue,
And emit beta particles too.
Your chemistry, still a mystery,
But we're learning more, it's true.
Promethium, oh Promethium,
You're not found in nature, it seems.
But in labs and reactors,
You're the stuff of scientists' dreams.

Samarium 62Sm

Samarium, a rare earth element so true,
Its atomic number is sixty-two,
Named after a Russian mine,
This metal is quite fine.
Soft and silvery is its hue,
It's used in magnets, did you knew?
Also in atomic clocks,
And lasers that tick like clocks.
It's in the periodic table's lanthanide row,
With elements like cerium and gadolinium in tow,
But samarium stands out,
With its unique magnetic clout.
So let's give a cheer,
For samarium, oh so dear,
A metal that's quite rare,
But in science, it's beyond compare.

Europium 63Eu

Europium, oh Europium,
Rare earth metal, so precious and rare,
In the periodic table, you sit,
Your properties, unique beyond compare.
You glow in the dark, like a star in the night,
A phosphorescent light, so soft and so bright,
You're used in TVs, computer screens,
A true innovation, so vivid and clean.
Your magnetic properties, so strong and so true,
Used in MRI machines, to see us through,
You're also used in nuclear reactors,
A power source, that's safe and secure.
Europium, oh Europium,
Your uses, so many, your value, so high,
A rare and precious metal,
In science, you'll never die.

Gadolinium 64Gd

Gadolinium, a rare earth metal bright,
In the periodic table shines with might,
Paramagnetic, it loves the magnetic field,
Used in MRI machines, a diagnostic shield.
Its electrons are prone to flip and spin,
Giving rise to a unique magnetic spin,
A property that makes it perfect,
For imaging tests and medical effect.
In nuclear reactors, it finds a use,
As a control rod, it keeps reactions diffuse,
With a melting point of over 1300 degrees,
Gadolinium is tough and hard to please.
So here's to Gadolinium, element 64,
A metal that has so much more,
From medical scans to nuclear power,
It's a versatile element, with great power.

Terbium 65Tb

Terbium, symbol Tb, a rare earth element
Named after a Swedish town, so eminent
Soft and silvery-white, it's easily cut
But in air, it oxidizes, forming a crust
Its unique green fluorescence doesn't go amiss
Used in color TV tubes, it's hard to dismiss
Magneto-optic recording, its claim to fame
Sensitive to magnetic fields, it's not the same
Terfenol-D, its alloy, is quite the wonder
Magnetostrictive material, it doesn't blunder
Expanding under a magnetic field, it's true
Used in sonar systems, it's a breakthrough
Terbium, oh Terbium, you're quite the element
Your properties and uses are truly relevant
From TV screens to sonar systems, you shine
A rare earth element, so precious and fine.

Dysprosium 66Dy

Dysprosium, rare and true,
A metal strong and bright in hue,
In magnets, lasers, and nuclear rods,
Its uses are many, its power awes.
From lighting up a city street,
To powering a fighter jet's heat,
Dysprosium's strength and stability,
Make it a key element of modern ability.
Its magnetic properties, oh so fine,
Make it essential in motors divine,
And in computer memories, it's a star,
Storing data near and far.
Dysprosium, oh how we need,
Your metallic might, your magnetic creed,
For in a world that's ever-changing,
Your consistency is quite amazing.

Holmium 67Ho

Holmium, oh holmium, a rare earth treasure,
In magnets and lasers, you shine with pleasure.
Your magnetic properties, oh so strong,
Make you a vital element all along.
In nuclear reactors, you stay stable,
Resisting radiation, you are able.
Your uses are many, your beauty rare,
In green and blue spectra, you shine without compare.
Holmium, oh holmium, you are quite unique,
Your atomic structure, a wonder to seek.
In medical imaging, you play a role,
As a contrast agent, you help achieve the goal.
In alloys and ceramics, you lend your might,
Making them stronger, more durable and bright.
Holmium, oh holmium, a gem of the earth,
Your properties and uses, of great worth.

Erbium 68Er

Erbium, oh erbium, with atomic number sixty-eight,
Your beauty, your strength, in modern tech we celebrate.
Your properties are unique, your uses oh so diverse,
In lasers, amplifiers, and data storage you immerse.
Your pink hue is enchanting, like a rose in the sun,
Your magnetic properties make us want to run,
To explore, to discover, to harness your power,
Erbium, oh erbium, you are our shining tower.
In fiber optic networks, you amplify the light,
In nuclear medicine, you aid in cancer's fight,
In metallurgy and alloys, you strengthen and enhance,
Erbium, oh erbium, your uses are vast and grand.
So here's to you, erbium, our versatile friend,
May your properties continue to guide and extend,
Our modern world, our technology, our future so bright,
Erbium, oh erbium, you are our guiding light.

Thulium 69Tm

Thulium, dear Thulium,
Rare earth and precious gem,
Soft and silver, yet strong,
In your properties we are drawn.
Your electrons dance in unison,
Creating your magnetic personality,
In lasers, you find your purpose,
Aiding in surgery with great clarity.
Your isotopes are used for dating,
Revealing secrets of the past,
In nuclear medicine, you're shining,
Aiding in diagnoses that last.
Thulium, dear Thulium,
Your uses are vast and varied,
In our modern world, you are vital,
A gem that will never be buried.

Ytterbium 70Yb

In the world of elements, rare and precious,
There lies one that's truly wondrous,
Ytterbium, with atomic number 70,
A gem that sparkles with metallic glow.
This element, so silvery and soft,
Is a treasure that's sought and oft,
Its properties, magnetic and unique,
Are used in science, for all to seek.
From surgery to nuclear medicine,
Ytterbium plays a vital role therein,
Its isotopes, used for dating rocks,
Reveal secrets of time that no one talks.
In lasers, it finds its place,
Creating beams with a steady pace,
And in atomic clocks, it's the star,
Keeping time like no other so far.
Ytterbium, a gem that's hard to find,
A rare earth with a brilliant mind,
In our modern world, it plays a part,
A treasure that's close to our heart.

Lutetium 71Lu

Lutetium, oh Lutetium,
Rare earth metal shining bright,
Symbol Lu, atomic number 71,
In the periodic table's sight.
From gadolinium's decay,
It's formed in nuclear reactors,
Used in cancer therapy,
And for detecting hidden factors.
Lutetium's magnetic personality,
Makes it useful in MRI machines,
It helps in surgery and nuclear medicine,
Aiding in procedures, so routine.
Its unique properties and uses,
Make it a valuable element indeed,
Lutetium, oh Lutetium,
A vital part of our modern world's need.

Hafnium 72Hf

In nature, it's quite rare,
But in technology, it's fair,
Hafnium is its name,
Its properties bring it fame.
Used in nuclear reactors,
And electronic factors,
Hafnium is quite versatile,
In its uses, it's agile.
It's resistant to corrosion,
And highly conductive with precision,
Hafnium is a metal,
That makes technology settle.
From aerospace to medicine,
Its uses are quite limitless within,
Hafnium is a crucial element,
In our world, it's quite prominent.

Tantalum 73Ta

Tantalum, oh Tantalum,
Your name inspires a sense of awe,
A metal that's rare and valuable,
With properties that leave us in awe.

You're strong and dense, resistant to corrosion,
Able to withstand high temperatures,
Used in electronic capacitors,
And surgical implants with great measures.

From cell phones to jet engines,
You play a crucial role,
Aiding in the advancement of technology,
With your remarkable properties so bold.

Tantalum, oh Tantalum,
A true wonder of the periodic table,
May your contributions continue,
As you help build a better world, stable.

Tungsten 74W

Tungsten, oh tungsten,
Strongest of metals, unmatched in its heft.
From filament to armor,
It's the backbone of technological progress.
Tungsten, oh tungsten,
Resistant to heat, impervious to wear.
Its strength is unyielding,
A force to be reckoned with, beyond compare.
Tungsten, oh tungsten,
Found in the earth, but forged by the hand.
A symbol of resilience,
Defying the odds, and taking a stand.
Tungsten, oh tungsten,
A marvel of chemistry, a wonder to behold.
A true testament to human ingenuity,
A shining example of the power of the bold.

Rhenium 75Re

Rhenium, oh Rhenium,
A metal rare and strong,
In alloys, it enhances,
And resists corrosion long.
A catalyst for reactions,
In jet engines, it's found,
With tungsten, it's alloyed,
To make filaments profound.
In X-ray tubes, it's vital,
To produce photons bright,
And in electrical contacts,
Its conductivity takes flight.
With one of the highest melting points,
And a density quite high,
Rhenium, oh Rhenium,
You're a precious ally.
In science and technology,
Your value is immense,
Rhenium, oh Rhenium,
You make our world advance.

Osmium 76Os

Osmium, oh Osmium,
Heavy and rare,
A metal so precious,
It shines with a glare.
With atomic number seventy-six,
It's one of the densest,
And it's used for many things,
From fountain pen tips to electrical contacts.
It's resistant to corrosion,
And it's hard as a rock,
A testament to its strength,
And the qualities it unlocks.
Osmium, oh Osmium,
A metal of great worth,
It may be rare and expensive,
But it's a treasure on Earth.

Iridium 77Ir

Iridium, oh Iridium,
A rare and precious metal,
In the Earth's crust, so little of them,
Found in places quite unsettled.
Its properties so unique,
A hard and dense material,
With a melting point so high and sleek,
It's used in aerospace materials.
In catalytic converters, it's a star,
Reducing pollutants from cars,
And in electronics, it goes far,
Making hard drives go faster and far.
Iridium, oh Iridium,
A wonder element indeed,
Used in science and technology,
For a better world to proceed.

Platinum 78Pt

Platinum, oh platinum, so rare and so pure,
A metal so precious, it's hard to procure.
With a lustrous white shine, it's a sight to behold,
And its uses so varied, worth more than gold.

From jewelry to medicine, platinum finds its way,
In catalytic converters, it helps clean the air each day.
Its resistance to corrosion, makes it last so long,
And in electronics, it's used to make them strong.

Platinum wires, electrodes, and contacts too,
In lab equipment, it helps experiments come through.
And in petroleum refining, it plays a key role,
Helping to extract usable fuels from crude oil.

So here's to platinum, a metal so fine,
A valuable element, with uses that shine.
May it continue to serve us, in ways both big and small,
For this precious metal, is worth its weight in gold after all.

Gold 79Au

Of all the elements that we hold dear,
None quite sparkles like the noble Gold.
Mined from the earth with sweat and tear,
Its lustrous sheen a treasure to behold.
A symbol of wealth, it's true,
But Gold has uses beyond measure.
From medicine to circuitry anew,
It's a vital part of modern treasure.
Its conductivity is unmatched,
A metal of choice for the pros.
In aerospace it's frequently dispatched,
Where its properties take us to new highs and lows.
And let's not forget its beauty,
A precious metal for all to see.
From jewelry to coins, it's a duty,
To showcase this element's majesty.
So let us raise a glass to Gold,
An element so versatile and bold.
May it continue to shine bright and old,
A symbol of value, worth and hold.

Mercury 80Hg

Mercury, oh Mercury, liquid silver so shiny,
A metal so fascinating, so fluid, so mighty.
At room temp, you're a liquid, a rare sight to behold,
But when it's cold, you solidify, and your beauty unfolds.

Your toxicity is well-known, yet we still use you with care,
In thermometers, switches, and lighting, you're everywhere.
You're used to extract gold and silver, and in dental amalgams too,
But we must handle you with caution, for you can be quite a brew.

Oh Mercury, oh Mercury, your symbol's Hg,
Named after the Greeks' swift messenger, how fitting indeed.
You're a heavy metal, with a density so high,
And your atomic number is 80, oh my!

From ancient times to present day, you've been a part of our lives,
A metal so versatile, yet dangerous, we can't deny.
But we'll continue to use you, with caution and with care,
Oh Mercury, oh Mercury, our liquid silver so rare.

Thallium 81Tl

Thallium, oh Thallium,
A metal that's rare and strange,
With a lustrous, silvery sheen,
Your properties are quite unique.
Soft and malleable you are,
But also quite toxic and bizarre,
A poison that can cause great harm,
Yet used in medicine as a charm.
Your atomic number is eighty-one,
And your weight is heavy as a ton,
You're found in minerals like crooksite,
And used in electronics with delight.
Your chemistry is quite complex,
And your properties are hard to vex,
You conduct heat and electricity,
And your melting point is quite nifty.
Oh Thallium, you may be rare,
But your uses are beyond compare,
From infrared detectors to green fireworks,
Your applications are quite diverse.
So here's to you, Thallium,
A metal that's strange and rare,
May your properties be explored,
And your secrets be laid bare.

Lead 82Pb

Lead, oh lead, heavy and dense
A metal with a toxic sense
Used for pipes and batteries alike
But its dangers we cannot strike
In the air, in the soil, in the water too
Lead pollution is a problem we must pursue
From brain damage to developmental delays
Lead exposure can have lifelong ways
Yet in pencils, lead is just a name
A harmless substance that brings us fame
For writing, drawing, and artistic expression
Lead brings joy without any aggression
So let's handle lead with utmost care
And protect ourselves from its toxic snare
For in moderation, lead can be a tool
But in excess, it can make us a fool.

Bismuth 83Bi

Bismuth, a metal of many colors,
With a rainbow sheen that never dulls,
From pink to blue and even yellow,
Its beauty shines, a sight to behold.
In medicine, it has its place,
As a treatment for stomach pain,
And in cosmetics, it adds grace,
To shimmer and shine, it's not in vain.
But in alloys, it's most renowned,
With lead and tin, it melts and blends,
Low-melting point, easy to work around,
Bismuth alloys, a crafter's friend.
So let us hail this element rare,
With its unique properties and flare,
Bismuth, oh how we do declare,
Your beauty and usefulness, we share.

Polonium 84Po

Polonium, oh Polonium,
A rare and deadly sight.
With alpha particles it decays,
And gives off gamma light.
Discovered by Madame Curie,
Its power she did see.
Used in atomic bombs,
And espionage, sadly.
But Polonium has its uses,
In static eliminators and heaters.
In isotopes for cancer treatment,
And in space as power generators.
A double-edged sword,
This element of the past.
Used for good or for evil,
Its power forever lasts.

Astatine 85At

Astatine, oh rare and fleeting,
Found in nature, but not in greeting.
One of the rarest elements on Earth,
Its beauty and uses, little of worth.

Radioactive, with a half-life so brief,
Its properties are hard to believe.
But in medicine, it finds its use,
Targeting cancer cells with precise abuse.

Astatine, a tool for research,
Leads to discoveries, a treasure to fetch.
In nuclear physics, its power is felt,
A tiny element, with a force to melt.

Oh Astatine, so little known,
Yet in science, its worth is shown.
A mystery, a wonder, a rare find,
May we unlock your secrets, oh element divine.

Radon 86Rn

Radon, noble gas of the night,
Inert and colorless, out of sight,
A radioactive element, so rare,
With atomic number 86, quite fair.
Born from the decay of uranium,
It seeps through rocks, a silent anthem,
A danger in homes, a risk to health,
A stealthy gas, a silent stealth.
But in medicine, it finds its place,
A source for treatment, a healing grace,
Used in radiotherapy, a ray of hope,
A chance for life, a way to cope.
So though it's fleeting, this element of fame,
Radon holds both good and evil claim,
A reminder that in chemistry's game,
Nothing is quite as simple or tame.

Francium 87Fr

Francium, elusive and rare,
A treasure to chemists who dare,
To seek out this element so fleeting,
Its properties so intriguing.
Its atomic number is eighty-seven,
And it's found in uranium's heaven,
But only in trace amounts,
In nature it hardly amounts.
Its uses are few and far between,
But in research, it's a valuable scene,
In studying heavy ion reactions,
Francium helps scientists make transactions.
Its half-life is just a few minutes,
But in that time, it shows its uniqueness,
In the periodic table, it's a marvel,
A testament to chemistry's marvel.
Francium, a mystery in its own right,
A fleeting element, yet shining bright,
In the world of chemistry, it's a gem,
A rare and precious chemical stem.

Radium 88Ra

Radium, oh Radium, your glow so bright,
A source of wonder, a source of fright.
With power to heal, with power to kill,
Your properties leave scientists still.
Marie and Pierre, they found you first,
In pitchblende, where you were dispersed.
Your radiation, so intense and strong,
Brought forth a new era, a scientific song.
But dangers lurk in your radioactive state,
Exposure to you can lead to a dire fate.
Cancer and burns, a deadly toll,
Caution is key, lest we lose control.
Yet still you're used, in treatments and more,
A paradox of danger and cure.
Radium, oh Radium, a wonder to behold,
In your mysteries, we'll forever be told.

Actinium 89Ac

Actinium, oh Actinium,
A rare earth metal, so precious and rare,
First found in pitchblende, a uranium ore,
Its properties, so unique and so fair.
With a shiny, silvery-white hue,
Actinium's beauty, so stunning and true,
Its melting point, so low and so few,
A radioactive element, so strong and so new.
Actinium, so useful in medicine,
A source of alpha particles, so serene,
Its isotopes, so varied and clean,
A breakthrough in cancer treatment, so keen.
Oh Actinium, a wonder to behold,
Its chemical properties, so rare and so bold,
A rare earth metal, so precious and so old,
Actinium, the element, so fascinating and so sold.

Thorium 90Th

Thorium, noble and strong,
In the earth, you do belong.
A metal, lustrous and bright,
Your properties, a wondrous sight.
Your atomic number, ninety,
Your uses, aplenty.
In gas mantles, you shone,
As a catalyst, you have shown.
Your isotopes, so diverse,
In fuel rods, you disperse.
But beware, oh mighty Thorium,
Your decay, a dangerous euphoria.
Your potential, immense,
As a clean energy source, you make sense.
Nuclear reactors, you power,
A sustainable future, you empower.
Thorium, a marvel of chemistry,
Both beneficial and risky.
May we use your power with care,
And handle you with utmost fare.

Protactinium 91Pa

Protactinium, a rare and precious find,
A metal that captivates the curious mind,
Its isotopes diverse, its nature strong,
In the depths of Earth, it stays for long.
A child of Uranium, it shares its fate,
A product of decay, but not a mere trait,
For its potential is great, its uses vast,
A source of energy, built to last.
But beware of its dangers, its decay intense,
Handle with care, and use common sense,
For a sustainable future, we must strive,
And Protactinium could help us thrive.
So let us explore, but with caution and care,
For this element, so rare and so rare,
May hold the key to a brighter tomorrow,
With Protactinium, we can overcome any sorrow.

Uranium 92U

Uranium, oh Uranium,
A metal of great power,
Your atomic might,
Can light up the darkest hour.
From the depths of the earth,
You were born in a supernova's blaze,
Your radioactive nature,
Is both a blessing and a craze.
Your isotopes are many,
Each with its own unique flair,
Some used in nuclear fuel,
While others are quite rare.
Your energy is harnessed,
In reactors around the world,
Your potential is vast,
Yet your dangers must be told.
Uranium, oh Uranium,
A metal of great debate,
May we use your power wisely,
And avoid a nuclear fate.

Neptunium 93Np

Of all the elements that we can see,
Neptunium is a mystery.
Named after the god of the sea,
It's a radioactive curiosity.
First discovered in a lab,
Its properties we still try to grab.
A key part of nuclear energy,
But it must be used responsibly.
It's a metal with a silver glow,
And its isotopes can be used to know
How reactors work and how they flow,
And how we can keep them from being foe.
But even with all its potential power,
Neptunium must be handled with care every hour.
For if we don't respect its radioactive might,
We could cause harm and see a dangerous sight.
So let us appreciate Neptunium's worth,
And use it to power our world with mirth.
But let us never forget the risks it brings,
And use it responsibly, in all our doings.

Plutonium 94Pu

In labs and reactors, you're found,
A metal, rare and dense, profound,
You're born in stars, a cosmic child,
But now on Earth, you're often reviled.
Your power is great, your energy vast,
But your radioactive nature is a danger to last,
You must be handled with utmost care,
For your toxicity could cause despair.
But in the right hands, you can do good,
A source of energy, misunderstood,
May we use you wisely, may we use you well,
And avoid the dangers of your radioactive spell.

Americium 95Am

A metal of curiosity and might,
Americium shines with radioactive light.
Synthesized in labs and atomic blasts,
This element's power is unsurpassed.
Its electrons dance with atomic glee,
Radiating energy for all to see.
But with great power comes great responsibility,
Handle with care, or face the penalty.
Americium, a symbol of science's might,
A beacon of progress shining bright.
Let us use its power wisely and well,
And in its energy, our future propel.

Curium 96Cm

Curium, oh curium, a radioactive wonder,
A man-made element, with power to thunder.
Nuclear reactors, and bombs it can fuel,
With power so great, it can make the earth drool.
Named after the Curies, a fitting tribute,
Their work with radium, a similar pursuit.
Curium's discovery, a feat of science,
With isotopes aplenty, and a half-life defiance.
In labs it's used, for research and more,
But its dangers are real, that we can't ignore.
Radioactive decay, with particles flying,
Can harm living things, and have us all crying.
So let's use curium, with caution and care,
Harness its power, but always beware.
For in the wrong hands, it can bring devastation,
A lesson we've learned, from history's translation.

Berkelium 97Bk

Berkelium, a rare element,
Heavy and radioactive, its power is evident.
Discovered in Berkeley, its namesake,
Its properties are still largely opaque.
A man-made element, it's not found in nature,
Its potential uses, still a subject for future.
Its isotopes are highly unstable,
And handling it requires great caution, it's no fable.
Used in research and nuclear science,
Its applications still in their infancy, no reliance.
Yet its power cannot be denied,
And with great responsibility comes a great stride.
Berkelium, a symbol of human ingenuity,
May its power be harnessed with great sagacity.
May we use it to unlock mysteries of science,
And not let its power be used for malevolent defiance.

Californium 98Cf

In the lab, a curious sight
Californium, radioactive light
A metal with a power so great
A force that we must regulate
Its uses, both for good and ill
A double-edged sword, a skill
A tool for cancer treatment, true
But also danger, we must construe
Handle with care, this element rare
Lest its power turn to despair
A cautionary tale, we must heed
For Californium, a potent seed.

Einsteinium 99Es

Einsteinium, a rare and powerful thing,
Named after a genius, it's fit for a king.
With 99 protons, it's not for the weak,
But in research and science, it's the answers we seek.
Born in stars, forged in exploding debris,
Element 99 is a sight to see.
Its uses are varied, its potential immense,
From nuclear medicine to isotopic defense.
But caution is key, for this element is hot,
And its radioactivity can't be forgot.
Handle with care, with gloves and with shield,
For Einsteinium's power must always be sealed.
So let's explore this element, with respect and with awe,
For its secrets and wonders are what we're here for.
Einsteinium, a rare and powerful thing,
In science and research, it's fit for a king.

Fermium 100Fm

Fermium, oh Fermium,
A fleeting and rare sight,
Born of a fission reaction,
In the darkness of the night.
Named for Enrico Fermi,
A pioneer of atomic power,
You hold within your nucleus,
A strong and potent flower.
Your half-life is short,
A mere flash in the pan,
But in that brief existence,
You show us all you can.
Your power is immense,
But with it comes great care,
For in the wrong hands,
You could cause great despair.
Fermium, oh Fermium,
A marvel of science and might,
May we always use your power,
In the service of what's right.

Mendelevium 101Md

Mendelevium, a fleeting dream,
A rare and fleeting element, it seems.
Named for Mendeleev, the father of the table,
But its existence is unstable.
Created in labs with great care,
For its power we must be aware.
With just a few atoms, it can emit
Radiation that could cause great detriment.
But with great power comes great responsibility,
To use it for good, with utmost sensibility.
In science and medicine, it holds great potential,
To benefit humanity, in ways exponential.
Mendelevium, a rare and fleeting wonder,
May its power be used with care, not plunder.

Nobelium 102No

Nobelium, a rare and fleeting friend,
A fleeting glimpse, we see its end.
Created in labs with efforts great,
Its properties we're yet to contemplate.
Named for a man of peace and change,
Its potential for good we can arrange.
A tool for science and medicine,
Its uses can be many and again.
But with great power comes great responsibility,
To use it wisely, with great ability.
For in the wrong hands, it can be a threat,
A weapon of destruction, we must not forget.
So let us strive to understand,
This element that's both rare and grand.
A symbol of human progress and might,
May we use it for good, and never for spite.

Lawrencium 103Lr

Lawrencium, a fleeting friend,
A rare and precious element,
With potential for good in science,
And medicine, a crucial element.
But with great power comes great risk,
Lawrencium, a potential threat,
If used improperly, a danger,
A cautionary tale, we must not forget.
Handle with care, this precious find,
Use wisely, for the good of humankind,
Unlock the secrets of the universe,
With Lawrencium, we can surely find.
So let us use this element with care,
And harness its power with great awareness,
For science, medicine, and humanity,
Lawrencium can be a true benefactor.

Rutherfordium 104Rf

Rutherfordium, oh Rutherfordium,
Element 104, discovered in our labium,
Named after the great physicist Ernest,
It's a radioactive metal, don't ingest.
With a short half-life of just an hour,
It's not useful for everyday power,
But in research, it has its place,
Helping us understand the atomic space.
Its properties are still not well-known,
But we see potential, like a gemstone,
For new discoveries and breakthroughs,
In fields like physics, chemistry, and bios.
So let's handle Rutherfordium with care,
Protecting ourselves from its radioactive glare,
And using it for the greater good,
Innovating science, as we should.

Dubnium 105Db

Dubnium, oh Dubnium,
A fleeting element, so rare
Created in a lab, with great care
Named after a city, with pride
In Russia, where it did reside
For only a moment, before it did decay
But in that moment, what did it say?
Dubnium, a metal so heavy
Its discovery, oh so heady
A fleeting glimpse of something new
A chance to explore, to pursue
What secrets do you hold, Dubnium?
What mysteries, what joys, what fun?
We may never know, but we'll persevere
For science, for knowledge, without any fear.

Seaborgium 106Sg

Seaborgium, element one zero six
A fleeting glimpse, a tantalizing fix
Synthesized in labs with expert care
To study its properties, rare and fair
Named for Glenn Seaborg, a man of science
Whose work and research led to its compliance
With the periodic table, a proud addition
To the elements, a noble tradition
A heavy metal, unstable and fleeting
Its half-life short, a fact worth repeating
But in its brief existence, it reveals
Secrets of the universe, how matter feels
Handle with care this precious element
Use it for good, for knowledge, in every experiment
For Seaborgium, like all elements, has a role
In understanding the universe as a whole.

Bohrium 107Bh

Bohrium, oh Bohrium,
Element one-two-seven,
So rare and fleeting,
A true wonder of heaven.

Synthesized in labs,
Created for research,
Its properties unknown,
Its secrets still a search.

Named after Niels Bohr,
Who saw the atom's core,
Bohrium is a tribute,
To his legacy forevermore.

In the universe so vast,
Bohrium is a speck,
But in the world of science,
Its discovery's a check.

A reminder of our limits,
And also of our might,
Bohrium is a symbol,
Of knowledge shining bright.

Hassium 108Hs

Hassium, a fleeting moment in time,
A wonder of science, so hard to find.
Discovered by chance, a fleeting flash,
In experiments bold, and moments brash.
Named for a land, once torn apart,
A testament to science, a brand new start.
A symbol of progress, of discovery's call,
Hassium stands tall, proud and tall.
Its properties unknown, its secrets untold,
Hassium beckons, a mystery to behold.
A fleeting moment, a glimpse of the new,
Hassium's story, forever anew.

Meitnerium 109Mt

Meitnerium, oh Meitnerium,
Rare and fleeting, like a dream.
Named for Lise Meitner, so wise,
Who unlocked the secrets of the atom's guise.
Your atomic number is one-hundred-and-nine,
And in your nucleus, protons align.
But you are so elusive, hard to find,
A fleeting glimpse, a moment in time.
In labs, they try to make you stay,
But you decay in a matter of days.
Your properties, still unknown,
But scientists keep searching, never alone.
Meitnerium, oh Meitnerium,
What mysteries do you still hold?
A key to the universe, untold.

Darmstadtium 110Ds

Darmstadtium, oh Darmstadtium,
A fleeting element, so elusive andrium,
Synthesized in labs, with great care,
Its properties, a mystery, so rare.
Named after the city, it calls home,
Its existence, still unknown to some,
With a half-life, so short and swift,
Its discovery, a remarkable gift.
Heavy, and metallic, in its form,
In the periodic table, it does conform,
To the group of transition metals, so true,
Darmstadtium, oh Darmstadtium, we marvel at you.
In the world of science, you hold a place,
A symbol of progress, in the human race,
And as we unravel, your secrets deep,
Darmstadtium, oh Darmstadtium, our curiosity, you do keep

Roentgenium 111Rg

Roentgenium, oh Roentgenium,
A fleeting element, elusive and rare,
Named after the man who discovered X-rays,
It's properties still shrouded in mystery and glare.
With an atomic number of one-hundred and eleven,
It's synthesis requires a nuclear fusion,
It's existence is brief, a mere fraction of a second,
Yet it holds great importance in scientific discussion.
Roentgenium, oh Roentgenium,
A testament to the wonders of chemistry,
A puzzle yet to be fully solved,
But with each discovery, we move closer to its mystery.

Copernicium 112Cn

In the depths of the periodic table
Lies a metal, rare and unstable
Its name is Copernicium, a mystery
With properties that remain a history
Named after the astronomer who showed
The sun, not the earth, to be the center of the code
Copernicium too, defies convention
Breaking the rules of atomic composition
Its isotopes, short-lived and fleeting
Are created through nuclear heating
Yet, despite its transience and rarity
Copernicium, to science, is a necessity
For in understanding its properties and traits
We unlock the secrets of the atomic gates
And pave the way for new discoveries
A future built on scientific recoveries
So let us marvel at this element's wonder
A symbol of human progress and thunder
For in Copernicium, we see the power
Of chemistry, in its finest hour.

Nihonium 113Nh

Nihonium, a fleeting fame
A synthetic element, without a name
Created in labs, through fusion's game
In the depths of science, it rose to claim
Atomic number 113, its identity
Named after Nihon, Japan's entity
A symbol Nh, of its rarity
A mystery, shrouded in anonymity
Unstable and fleeting, its existence
A fleeting moment, with persistence
A contribution to science's essence
A discovery, with significance immense
Nihonium, a wonder of chemistry
A fleeting element, of the unknown mystery
A symbol of progress, and human ingenuity
A rare gem, in the periodic table's flurry.

Flerovium 114Fl

Flerovium, oh Flerovium,
A synthetic element so rare,
Created in a lab, not in a medium,
A testament to human's scientific flair.
Named after a great scientist,
Who paved the way for nuclear physics,
Flerovium's existence is so distinct,
It's a symbol of modern chemistry's tricks.
With just a fleeting lifespan,
Flerovium may seem insignificant,
But it's a product of human's plan,
To unlock the secrets of the periodic element.
So let us celebrate this element,
For its rarity and significance,
And let it remind us of the moment,
When human's curiosity leads to scientific brilliance.

Moscovium 115Mc

Moscovium, the element of wonder,
A fleeting existence, a cosmic thunder.
Created in labs, by human hand,
It's a testament to science's command.

Named after Moscow, the land of its birth,
It's a symbol of progress, of scientific worth.
A heavy element, with an atomic weight,
It's a wonder of nature, a scientific fate.

Its properties are unknown, a mystery to solve,
A puzzle for scientists, to constantly evolve.
Yet, it stands tall, a symbol of human curiosity,
A testament to science, and its boundless creativity.

Moscovium, a fleeting beauty,
A product of human curiosity.
A wonder of science, a mystery to solve,
An element of the future, for scientists to evolve.

Livermorium 116Lv

In the depths of the periodic table's maze,
A new element was found, to our amaze.
Livermorium, a synthetic creation,
A fleeting existence, a rare sensation.
Named after the city where it was made,
Livermore, California, where it played.
A heavy element, with 116 protons,
Once discovered, it sparked a commotion.
Its properties, still not well-known,
But its discovery, a milestone.
A testament to human curiosity,
And the wonders of scientific ingenuity.
Livermorium, a tiny piece of matter,
But its discovery, a scientific chatter.
A reminder that there's still much to learn,
In the vast and complex world we yearn.

Tennessine 117Ts

In the depths of science labs,
Where curiosity collides,
A new element was born,
With protons to divide.
Tennessine, the name it bears,
A testament to human quest,
To unravel nature's mysteries,
And put its secrets to the test.
In Oak Ridge, Tennessee,
The element was first made,
A fusion of atoms, a breakthrough,
A discovery that won't fade.
Named after the state it was found,
Tennessine is rare, it's true,
But its significance is profound,
A symbol of what we can do.
For science is a journey,
A never-ending quest,
To unlock the secrets of the universe,
And put our knowledge to the test.
So here's to Tennessine,
And all the elements we've found,
May they inspire us to keep exploring,
And uncovering nature's profound.

Oganesson 118 Og

Oganesson, a fleeting form,
A synthetic element, born to transform.
Named for Yuri Oganessian,
Whose team found this element, a precious lesson.
118 protons in its core,
This noble gas, a mystery to explore.
Unstable, heavy, and rare,
It defies the laws of chemistry with flair.
Its half-life is incredibly brief,
A mere fraction of a second, a time so brief.
But in that moment, it shines,
A symbol of science, pushing boundaries, and lines.
Oganesson, a wonder to behold,
A testament to human curiosity, bold.
A reminder that the universe is vast,
And the secrets it holds, we'll uncover at last.

Periodic Table of Elements

1 H																	2 He
3 Li	4 Be											5 B	6 C	7 N	8 O	9 F	10 Ne
11 Na	12 Mg											13 Al	14 Si	15 P	16 S	17 Cl	18 Ar
19 K	20 Ca	21 Sc	22 Ti	23 V	24 Cr	25 Mn	26 Fe	27 Co	28 Ni	29 Cu	30 Zn	31 Ga	32 Ge	33 As	34 Se	35 Br	36 Kr
37 Rb	38 Sr	39 Y	40 Zr	41 Nb	42 Mo	43 Tc	44 Ru	45 Rh	46 Pd	47 Ag	48 Cd	49 In	50 Sn	51 Sb	52 Te	53 I	54 Xe
55 Cs	56 Ba	*	72 Hf	73 Ta	74 W	75 Re	76 Os	77 Ir	78 Pt	79 Au	80 Hg	81 Tl	82 Pb	83 Bi	84 Po	85 At	86 Rn
87 Fr	88 Ra	**	104 Rf	105 Db	106 Sg	107 Bh	108 Hs	109 Mt	110 Ds	111 Rg	112 Cn	113 Nh	114 Fl	115 Mc	116 Lv	117 Ts	118 Og

1	Atomic Number
H	Symbol
Hydrogen	Name

*	57 La	58 Ce	59 Pr	60 Nd	61 Pm	62 Sm	63 Eu	64 Gd	65 Tb	66 Dy	67 Ho	68 Er	69 Tm	70 Yb	71 Lu
**	89 Ac	90 Th	91 Pa	92 U	93 Np	94 Pu	95 Am	96 Cm	97 Bk	98 Cf	99 Es	100 Fm	101 Md	102 No	103 Lr

About the Author

Walter the Educator is one of the pseudonyms for Walter Anderson. Formally educated in Chemistry, Business, and Education, he is an educator, an author, a diverse entrepreneur, and is the son of a disabled war veteran. "Walter the Educator" shares his time between educating and creating. He holds interests and owns several creative projects that entertain, enlighten, enhance, and educate, hoping to inspire and motivate you.

WaltertheEducator.com